A Life of Perfume

A Collection of Essays by
A Master Perfumer

EDMOND ROUDNITSKA

Published by Leonard Payne

September 2018
England

ISBN 978-0-244-71322-5

Table of Contents

INTRODUCTION

I first came across Edmond Roudnitska's (ER) written work when I was searching for his essay on "The Novice and His Perfume Palette." It was hard to come by and was embedded in a badly scanned pdf file.

I later discovered that although ER was said to be quite prolific in his writings, most of those I came across were in French and there wasn't a lot of English language material. (I've since taken the liberty of translating some of the French material).

It's often been said that once something is published on the net, it stays forever. I've not found that to be true. I've found that sometimes authors who transgress some of the niceties of the social justice warrior mobs, tend to have their whole corpus of literature go missing. So in order to ensure that ER's material gets a wider audience and doesn't get 'lost' for some reason, I've privately published these. I hope they are found useful.

ER, whilst probably one of the most brilliant perfumers ever, was not without his 'feet of clay'. If it wasn't French it was not good. In the (paraphrased) words of the ever flamboyant Luca Turin, "French arrogance is fine in homeopathic doses…... Speaking of Turin, I wonder whether he got some of his ideas about the structure of odour being vibration rather than shape from ER? (See "Where are we going?")

Enjoy ---- Leonard Payne – September 2018

PORTRAIT OF A PERFUME COMPOSER

By Jean-Claude Ellena

Edmond Roudnitska has dedicated his life to the service of perfume. This portrait, presents it between 1935 and 1956. 20 years during which the man evolves and the perfume composer poses the principles of his art.

Paris, 1935. Edmond Roudnitska has the look and bitter face of an athlete. Tall and thin, he moves with a sure step. The voice is put. The look is safe and penetrating. The nose is aquiline. He is aware of his charm. He will be photographed by the studio

Edmond Roudnitska

Harcourt. From his father, son of Russian officer, he inherited elegance, Slavic temperament and fiery character. Hard for himself, he is demanding with others. If his passion for sport is pronounced, he has little interest in culture.

At the age of 30, Edmond Roudnitska has just signed a contract of *odor composer* with the company Fabriques de Laire, located in Issy les Moulineaux. This contract provides for the realization of compositions which he will send samples accompanied by broad and precise explanations: motivation, idea-support of perfume, description, proposal of presentation to the customers.

Since the beginning of the century, Laire factories have been internationally renowned for creating bases for perfumery, such as Amber 83, Bouvardia and Saxon Moss, as well as products of synthesis, such as hydroxycitronnellal, heliotropine and vanillin.

For Edmond Roudnitska, the knowledge of the composition was acquired 7 years ago in the company Roure and Justin Dupont in Argenteuil. Although he did his apprenticeship a few years in Grasse, in the raw materials analysis laboratory, it was during these three months of intense work in 1926, with Mr. Leon, a seasoned perfumer, that he acquired the technique of the composition, the spirit of the bases and decided of its vocation: to be perfumer!

In the 1930s, perfume chemistry provided perfumers with synthetic substances that are considered difficult to use with their violent, harsh, unadorned smells. The olfactory reference is the product of natural origin. Some perfumers specialize in the creation of bases or compositions of a dozen components in which a characteristic body of synthesis is tamed. This "broad strokes" formulation technique develops a synthetic but also analytical and rational mind. ER will compose many bases.

1938. He participates in the writing of the leaflet of the perfumery products of the Fabrique de Laire, 165 pages of explanatory notes, condensed and precise. This catalog reveals a rigorous and scientific spirit, it tells us that Edmond Roudnitska is already familiar with hundreds of bases and many chemicals such as methylionones for which he has a perfect mastery. Mastery envied by many fragrance composer still today.

More than 50 years later in "The Art of Composing", published in 1991 in "Art, Science and Technologies," the materials of his "keyboard" as he liked to describe his perfume organ, will be largely the natural essences and the synthesized bodies dating from his composer debut in Laire society Befriended with the young François de Laire, then director of commercial services, he meets the Parisian world. Elegant medium he had seen, child, in the salons of the prefecture of Nice where his parents worked. We talk about painting, music, literature, beauty. This "broth of culture" marks a turning point in his life, because he glimpses answers to his passionate job. He reads and re-reads philosophers and thinkers: Montaigne and Montesqieu. Discover the books of Étienne Souriau, Henri Delacroix and fascinates Henri Bergson.

1938 is also the year of his first article, published in the magazine "Parfums de France", the subject: fixers. If the reflections are essentially technical, Edmond Roudnitska denounces the abusive use of stubborn products, contrary to the criteria of compositions of perfumes of the time and rejects the part of chance in the composition of a perfume: *"a perfume modern is not a simple assembly of spare parts, but a whole, a whole, and it is necessary to leave it to another master only chance, to coordinate its elements."*

The man is encamped: intuitive and passionate, rational and deductive.

In 1942 he meets Therese Delveaux who will become his companion. She is young, brilliant chemical engineer and passionate about fashion. She works in the pharmaceutical sector of Laire's company. As the laboratories face each other, they meet, meet and share their inclinations. At her touch she begins to smell. He offers him his first creation "It's you" that he composed for Elizabeth Arden in 1938. Together they take long bike rides in the forest of Rambouillet or Isle-Adam to pick lily of the valley. In search of rare editions, they run the bookstores of Paris. Passionate about opera, he invites him to Palais de Chaillot to listen to Tannhauser conducted by Wilhem Furtwangler. Wagner ignites him.

At this same time he finds in Kant's reading the confirmation of his intuitions, when he encourages to " *think for himself* " and adds: *"to think is not something innate, that can be learned, driven according to rules, refined* ". Like Kant he favors intuition and increases " *intuition is cultivated, like taste, in contact with other arts and if there is no method to compose, there are methods of work".*

From 1945 on, he undertakes to synthesize the different critical essays of Kant 's "Aesthetics" and begins writing his own theories. Principles it will deliver in "Aesthetics in Question" 31 years later.

To give meaning to the practice of his profession of perfume composer, to release an aesthetic theory and a philosophy of life, these are his goals.

Time of war, time of scarcity. The scarcity of raw materials in all areas calls for imagination and creativity. While the formulas of its bases are written in large scale, less than a dozen raw materials, the formulas of its perfumes are complex: dozens of components, many bases, entangle, mingle, blend. But man is daring in using rebellious constituents and talent by using byproducts like the mythical "was V" body, a mixture of methylionone distillation bottoms with a pyrogenic odor.

In the 4O's, the fashion of women's perfumes is "cyprus", olfactory construction around woody notes, patchouli, animal notes, mosses and leather. Coty Cyprus, Lanvin Scandal, Shocking of Schiaparalli, Bandit de Piguet are the prototypes of these opulent and sensual feminine perfumes.

The perfume composer ER is in keeping with his time.

He composes for Marcel Rochas that he has just met, Femme, a fruity chypre. *In 1943 during the occupation, I freely composed the perfume "Femme" that Marcel Rochas launched as is underwriting in 1944, immediately after the liberation. This is not a Cyprus, as we tend to catalog, because most perfumes contain oak moss today, but are not Cyprus. "Femme" is an aldehydic floral perfume, very fruity,*

with a double woody and candied characteristic. This is his summary profile that is not that of a Cyprus and even less that of Coty's prototype".

The man does not want to be classified, he wants to be unique!

In 1946 ER set up his workshop in Becon-les-Bruyères, in the suburbs of Paris and created the company Art et Parfum. Fame is acquired he can devote himself intellectually to his art and develop his aesthetic approach.

To compose, to assemble, to order, to dose, to note, to give meaning to his compositions. The perfume is above all the creation of the spirit!

Perfumes and successes follow each other: Chiffon de Rochas (1946) which will be renamed Mousseline then, Fly, the Rose of Rochas (1948), and on an idea of Therese Roudnitska, Mustache of Rochas (1948).

At the same time, he met Serge Heftler-Louiche director of perfumes Christian Dior, and composed Diorama in 1947.

What a debauch of inventiveness in this perfume! Never perfume will have more complex form and formula, more feminine contours, sensual, carnal. Bases and products. It seduces us with its spicy notes: pepper, clove, cinnamon, nutmeg and cumin, with the smell of skin. It disturbs us with its animal notes: castoreum, civet and musk. All the chords and themes to come are contained in this perfume: the woody and the violet, the plum and the peach, the jasmine and the spices.

Cabris , 1949 . Edmond Roudnitska and his companion settle on the heights of Speracedes, near the village of Cabris in the Alpes Maritimes. Like all men who dream of human fulfillment, his gaze needs to focus on the source of our culture: the Mediterranean. The Mediterranean *These places where men had the sense of the right proportions and spoke a better language,* to quote Fernand Braudel. The decor is wild and dominates the sea. The property is only rocks and scrubland. The water will be captured below several hundred meters, with difficulty and at great expense.

The place deserves!

With its ionic columns and its inner garden the house, turned towards the sea, takes on the air of a Greek villa. When the east wind blows ER detects the smells of the sea, in southern wind those of the mimosas that cover the Tanneron.

Six hectares of Italian gardens, a wide variety of flowers, shrubs and trees, numerous paths, paths, and sculptures punctuate the landscape. At the threshold of the property is carved in stone: "I will make the stones bloom and sing the birds"

The place concretizes his thoughts. He can reflect, write, transmit these aesthetic principles. The articles will be numerous and will follow one another at regular intervals, especially since 1959.

In May 1951, at the conference: "The Music of Perfumes", made under the auspices of the Union of Technical Perfumery and Cosmetics, Étienne Souriau, professor of aesthetics at the Sorbonne, proposes the basic conditions necessary for an art of perfume to be introduced and he adds " *do not hesitate to recognize that the sense of smell is an aesthetic sense in the highest degree, and that it presents as much and better than any other sense, all who can to serve to support a complete art* "ER who is in the audience is particularly sensitive to this argumentation.

If the comparisons between the composition of a perfume and the musical composition proposed by Étienne Souriau do not suit him, Edmond Roudnitska finds here the springboard of his dialectic, and proposes the bases of an 8th art, in a conference ** that he gives in Paris on November 20, 1952, where he concludes: *for this to be an art, it is necessary that the sense of smell cease to be a meaning to be satisfied in order to become only a means.* Thus perfumes will be compositions of the mind and the public will be able to learn about olfactory forms.

The formulas of his perfumes are simplified. He gradually abandons the heavy, sweet notes. Renounce those tempting fruits and greedy like vanilla. Decides to compose with more rigor and privilege the pure, strictly olfactory notes. It was the creation of l'Eau d'Hermès in 1951 and L'Eau Fraîche in 1955 for Dior, the only command directed by Serge Heftler-Louiche who wanted a creation inspired by Coty's Cyprus.

Then he creates Diorissimo for Dior in 1956. At the opening of the bottle, green notes, bright, fresh, crumpled foliage you hang. An elegant wake, clear, limpid invades us. The theme: an armful of lily of the valley. The perfume is offered, we intoxicate, surprises us, plays jasmine, orange blossom, rose and lilac. There is in the game of these components a deep and shared pleasure of strand of lily of the valley picked, felt at dawn and breathed until evening. From now on, the perfumes he creates will be airy, sparse, fluid and clear like Diorrissimo. For Edmond Roudnitska " *Diorissimo is a return to Nature but, above all, reflects in its author, thirty years after its beginnings,* Diorissimo by its shape is also the witness of an ethics and an aesthetic research beyond the pleasant and aphrodisiac pleasure of perfumes. He will write some time later about his creations: " *talk to the public of form and not of epidermis".* The perfume composer has taken over the man. Freed from his emotions, his sense of smell became a tool. He adds *"I form a body with the perfume, the perfume is a part of me, I am a smelling machine"*

With the computer utopia of the sixties, he dreams of an art that has become measurable or " *well-designed machines, immune to colds, errors of judgment and pressures of all kinds, will take care of the choice of materials. first ... but also the synthesis of new bodies and the reconstituion evocative of natural scents, more faithful.*

With many successes, where each creation is the story of an intimate relationship between nature and him, composed in a refined style that avoids the commonplaces and facilities of a fashion, he devotes all his energy to writing several articles , then books, between 1974 and 1977, "The Intimacy of the Perfume", "To train the men, myth or reality?", "The

Aesthetics in Question" then "The Perfume" in the collection "What do I know? "because for ER, composing and writing is progressing in understanding nature and people. He knows the extreme difficulty of this task. All the more reason not to be distracted and to mobilize all your strength.

He will compose and write until the end of his life, at the age of 91.

Jean Claude Ellena 5/99

WHERE ARE WE GOING?

By Edmond Roudnitska – 1969

THE importance, among other things, of the organisation you wish to set up, is that it would at last fill the gap left by the dis-appearance of the work sections which we attended during the 'forties within the Groupement Technique de la Parfumerie.

Research workers of all kinds, both in this country and in countries where similar bodies might be created, would thus have a permanent meeting ground. I have in mind in particular the scicntists concerned with the theory of odour, who might thus be able to continue their efforts more easily.

This theory of odour, while it may not solve all our problems, will enable us to tackle some very important problems that are to-day out of reach. for instance, as regards the definition of the attributes of an odour and the corresponding units of measurement. Given that as a basis, one could begin to treat the subject rationally.

Many precious years have been wasted, during which scientists have taken little interest in this research. Those who might have been qualified to conduct it satisfactorily have dis-appeared. The sense of smell was regarded as a secondary sense, just as if anything concerned with life could be secondary, and as if one discovery did not lead to a series of others.

THEORY OF ODOUR

One of your tasks should be to stimulate energy and worry the researchers, chemists and above all physicists until the theory of odour is finally elaborated.

These questions are unfortunately outside my sphere of competence, and it is therefore only timidly that I place before you some reflections of mine on certain work by Louis de Broglie and the quantum theory.

I recapitulate:

It occurred to Louis de Broglie to allot a mass to the photon. The idea has been debated.

Light contains photons, it is a wave, and they must therefore co-exist.

Light and matter are forms of energy. According to Planck, energy equals *h* times the frequency, whence the sequence: Particles of matter have mass. Mass is a form of energy.

Energy implies frequency.

Frequency implies vibrations. The result of this is that the particles are endowed with vibra-tions.

Particles endowed with vibrations resemble photons.

Photons bear a relationship to light waves.

Therefore, matter should have a relationship with the "waves of matter".

Einstein has shown that light, which was for a long time conceived as a wave, resembled a particle. Broglie went full circle by suggesting that matter, for a long time considered as made of particles, must be accompanied by waves and therefore partake somewhat of their nature. Einstein ap-proved.

Is it absurd to imagine something similar for odour? Even if odour consists of particles of matter, may it not be thought that this particle can display behaviour similar to that of photons and give rise to vibrations?

If it has been possible for radiation to be transformed into matter, according to Einstein's law of the equivalence of mass and energy, it seems more plausible for matter to be transformed into radiation. Odour could be an aspect of a transfer of energy from matter. I confine myself to submitting this reflection to those more qualified to consider it.

MATERIALS AND THEIR SUPPLY

From a practical point of view, however, where are we going at the start of this last third of the century?

Chemical synthesis continues its progress in a particularly dazzling way. The elaboration of substances to which great importance attaches continues at accelerated pace; the development is more difficult to follow since the cost prices are sometimes high. A great effort should be made by organic chemists to improve their processes and the cost of their syntheses, if they wish to bring them within our reach and not confine themselves to theoretical research.

The producers of Jasmin who have just gone through quite a foreseeable crisis, after excesses that we have always denounced, have received the just need of their weakness in following the bad shepherds.

The return to normal should restore some sanity lo the situation and if, as we have urged, prices are stabilised at a reasonable level for several years, perfumers will again be encouraged to make wider use of jasmin in their formulae. It will then become possible to conclude between perfumers and producer's contracts covering 3, 4 or 5 years and to ensure price stability to the former and guaranteed markets to the latter.

There is not, and never has been, an over-production of Jasmin, but there has been an under-consumption as a result of prohibitive prices. There· have also been bad Jasmins. May those who have embarked on fresh plantations continue unruffled; if the price remains reasonable and the quality as it should be, Jasmin will sell well and its use will develop.

Some producers are boldly deciding to follow our advice on growing and treatment. The results obtained are so convincing that those who have not lent themselves to the same disciplines will be outdistanced in canvassing a market which is more and more open to competition.

'The improvement thus obtained in quality will allow and justify transfers (carry-overs) of harvests from year to year, which transfers must be kept in cold store, stainless packing, under nitrogen. An intelligent inventory of world needs would enable producers to allow a certain harvest surplus which, if properly stored, would act as a regulating reserve and would further promote price stability. Stabilisation of this kind has in fact taken place this year with Rose Maroc (Moroccan Rose), which in spite of reduced production has been able to maintain its price, thanks to the surplus from last year.

All these efforts take the form of a new contribution of outstanding pro-ducts, whose influence will make itself doubly felt: first of all by sharpening the sense of smell and the minds of the creators, to whom finer models are offered; next, by enriching their palette and opening up horizons that had hitherto been closed.

PERFUME QUALITY AND THE FUTURE

Generally speaking, how will the next few decades shape themselves?

With the scientific and technical revolution, which confirms the right of all to live in contact with beauty, it is to be expected that the public will display an increasing appetite for beautiful things, which will end by extending to all products on the market. In collective civilization, in which we are already involved, happiness implies aesthetic compensations.

However, there is a risk of aesthetics losing in quality what it will gain in quantity. 'To a certain extent, the high price of quality, by sustaining the religion of the beautiful among an elite of fastidious connoisseurs, op-poses a barrier to bad taste, while popularisation, by exceeding its aims, sometimes leads to vulgarity. Nevertheless, the beautiful is not necessarily expensive and yields a large return, simply because it is generally the sign of success.

It is because the beautiful is one of France's traditional exports that we must have the courage to eliminate the merchants of what is mediocre. What is more, Foundations should be encouraged by tax exemptions.

Furthermore, aesthetics must be given the place which it deserves in modern education. Education in all its domains-aesthetics in particular should be the major care of those responsible for training people.

It is said that we are moving towards an audio-visual civilisation. It is more likely that we are moving towards a sensorial civilisation in which perfume should be able to play a large part. Perhaps films and lectures will be accompanied by olfactory demonstrations. Will it one day be feasible, as has been done for images, to transmit odours through space? This would be the ideal solution for perfuming shows and spectacles, the odours ceasing to be perceived as soon as transmission has ceased.

In antiquity, perfumes impregnated everything. In a few years we may witness the same frenzy. There is therefore no occasion for anxiety in connection with the perfumery industry as such or for worrying over the future of the "odour merchants". However, perfume itself is threatened; it is more than threatened, it is attacked. Thus, instead of asking whether perfumery is still an art, it would be better to give it every chance of showing that it can still be one. The cinema, which was regarded as an industry, has changed to become an art. Television is moving in the same direction.

When our friend Billot wished for perfumery "a place where perfumes can be made for their own sake, without any other consideration", he knew that there is at least one such place, whence scarcely one perfume emerges every ten years, and where only what is thought right is done, without any commercialism. This method of approach nevertheless proves very remunerative, since the well-known House which applies it is thriving, and its use of this method or formula is often quoted as an example of good management. Those who are only concerned with profit should meditate on this modest example.

THE ROOTS OF INNOVATION

In 1962, I did not allow ten years for the new wave to become brilliantly manifest, and I still hope that it will not give me the lie. However, he who succeeds is not likely to be one who has had ideas instilled into him by others, or even one who, in search of profit, has sensed where the wind is coming from. It will be he who, without looking over his shoulder, as McLuhan says, has a really personal idea, will maintain it until some work is completed, and will remain himself proof against wind and tide.

I do not think it is the creative capacity that is lacking, or the capacity for understanding on the part of the public (which remains what it has always been). What is in process of petering out is the capacity for starting something, it is really responsible people who are lacking, and it seems that to-day there is no man capable of launching a Chanel No. 5. Here is the problem of perfumery, a problem of giving orders, a problem of authority. Above all else, this is the problem that must be tackled, and the rest will follow. However, authority is nothing without competence, and in a profession such as ours competence is not acquired overnight.

I therefore do not see any immediate way out. It is urgently indispensable for an animator, such as Coty was in his day, to come forward, stand out, and be listened to. His task will be the more difficult, since he will not be on virgin soil; he will have to eliminate a considerable number of parasites which occur in different forms. He will require much clarity of mind, much courage, energy and self-sacrifice. A kind of white blackbird! Who will discover him and persuade him to pick up a torch which has been dropped for so many years? For, through some kind of fatality, transfers take place with difficulty from generation to genera-tion in our profession. This is not only true of perfumery in its final state, but also of the raw materials industry. The number of abandoned items is on the increase.

Whereas in the motor car industry, for instance, in spite of the disappear-ance of the Citroen and the Renault, and in spite of pitiless competition, those makes are more prosperous than ever. In the case of Peugeot, quality has never been raised to so high a degree. In those large firms, the reins have most felicitously been placed in the hands of capable and conscientious managers who are capable not only of maintaining the reputation of their makes, but, enhancing them still further.

Why is not this the case in our industry? Doubtless because per-fumery is essentially based on taste, and that to be successful other qualities arc needed than those of skilled engineers and clever businessmen. It is high time that those financially responsible should take note and learn the art of discovering and then using the available skills (which implies that they must alter their criteria in the recruitment of top staff), if they do not wish to see the gradual crumbling of what has been the apanage of our country.

In cosmetics, which is a tech-nologist's industry based on science, it is possible to succeed with technical staff, by applying business methods, Perfume, however, is not just based on technique; it aims at expressing beauty. Its ambition is not to be useful, like a cream, but to please. In these two activities the design of the products is quite different, just as the reasons for purchase are different. It is not possible to manage these two industries with the same mentality, with the same methods, with the same men. If this is not realised, the splendid perfumery industry will be finally condemned and this will be the death of the goose that laid the golden eggs, whose sole desire is to lay them.

NOT MADE TO ORDER

It is said that innovation is the modern form of competition. If this formula is true for the products of industry, we must be careful not to apply it blindly to perfumery.

A perfume is not a practical object, it is a work which, not being merely subject to reason, has required research of an aesthetic order. Experience shows that true innovation in this domain is arduous and often a burden because it is laborious, and its acceptance by the public needs a period of adaptation, sometimes even after a period of rejection.

These factors curb the rhythm of bringing out perfumes, It is not possible to accelerate this rhythm without falling into the slough of contingencies and tiring the public's patience by repeated failures. This is what happens.

Jacques Maisonrouge says that "a product is arrived at in two ways. On the one hand the discoveries made in the domain of technology, which make it possible to carry out certain functions; and on the other the needs expressed by a certain market, showing that it must be possible to solve such and such a problem or meet such and such a need on the part of customers, This synthesis that is necessary be-tween the market demand and the possibility of technological achievements is one of the salient points of modern industry. It seems (he adds) that quite often in Europe, in the presence of new technological ideas, commercial services are quickly consulted which, not having very powerful market research services, reject or accept the product according to the intuition of the departmental manager."

The conflict with us is born of the confusion between these two roads. "Mass" perfumery might attempt to take the second road (provided that the market is capable of expressing its needs), but "high grade perfumery" owes it to itself to take the first road. And when quality products are really concerned, this first road leads to world success and finally to large scale production. Such success, if well exploited, can even end in mass production.

Cases are known in which market research services have rejected high grade products which, taking their chance elsewhere, have later succeeded.

They have also granted their favours to notoriously worthless objects, which the referendum had not set on one side and which constituted as many setbacks. In the present circumstances, it is not possible to use scientific foresight in the realm of aesthetics.

In order to be able to create a really new perfume, we must not be stampeded either by time or by men. This is not done to order. It sometimes takes years just to collect the necessary materials. The time must also be allowed for many repentances!

It is all these aspects that show the difference between creative perfumery and industrial perfumery. The former represents an innovation, while the second seldom does so. Heads of firms, who do not understand this and think that by increasing the attempts they automatically increase their chances, are acting against their interests and do wrong to the whole profession.

WITHOUT PREJUDICE

It may be the role of this academy to provide those responsible for perfumery with information devoid of all prejudice and to present it with courage and mature reflection. It could constitute a medium in which free opinions would be freely expressed and debated. It would thus become a living and fruitful centre, not confined or limited to merely academic debates.

CONCERNING THE AUTHOR

"In 1926, at the age of 21, Edmond-Fernand Roudnitska entered perfumery quite by chance. He started in Grasse, the perfume city, and then obtained employment with de Laire. where he remained a certain number of years.

"Until 1949 he remained at Paris making bases, composing perfumes, and showing himself to be a talented perfumer.

"As in the case of many other perfumers, it is impossible to speak of his outstanding or sensational creations without raising violent reactions on the part of those using them.

"In 1946 Roudnitska became his own master by establishing 'Art et Parfum', installed at Cabris. There he composes perfumes. He sells to users the compounded oils of his own creation. These customers are commercial organisations, perfumers or dress designers who make up his products in solution, look after packaging, or approach manufacturing works or packagers themselves, organising publicity and then sales.

"It will be appreciated that, in these circumstances, it is impossible to enumerate Roudnitska's achievements.

"Independently of his actual creations, Roudnitska has an active interest in the Ste. Technique des Parfumeurs de France. He has very often taken part in discussions, when that society was known as the Groupement Technique de la Parfumerie, always safeguarding the interests of our profession and helping to reveal its true artistic character.

"He published, in the journal Parfumerie, in 1944, a very interesting paper on the perceptibility of odours. Prior to this, in the Revue des Marques (now defunct) he had written some highly interesting technical articles.

"A few years ago he suggested an original and interesting solution for stabilizing the jasmin market, but unfortunately nothing came of this.

"Roudnitska is a truly creative perfumer in the fullest meaning of the word."-Marcel Billot, "A Short History of the Great Perfumers," Soap, Perfumery & Cosmetics, June 1962,

THE NOVICE AND HIS PERFUME PALETTE

Edmond Roudnitska

The following advice is derived from a lifetime of carefully reviewed experience. The knowledge on which it is based is of such vital importance in the proper development of a creator of perfumes, that the Author had up till now reserved this advice for the members of his family and a few close friends.

In the light of an uncertain future and in order that at least a minute portion of his dearly acquired knowledge should not be lost forever, the Author has decided to publish it.

The conditions under which a young perfumer first makes the acquaintance of odorants are of great importance to his learning process and constitute a major influence on his approach to his materials throughout his career.

A good way of teaching him to fix the smells in his memory is to group them in 'families" or "series" With the young people in whose training I had a hand, I limited myself to 15 series-citrus, rose, orange-flower, Jasmine, violet/iris, anise, aromatic, green, spices, wood, tobacco, fruit, balsamic, animal, and leather. This sequence of listing is deliberate. It starts with the families which are at once the simplest, the commonest and the broadest--like citrus and the characteristic floral notes-and moves via more specialized series to end up with the heaviest and most exciting.

There is no point in making the list longer; this would impair both the student's concentration and the "families" concept. So I see no reason to have a series for each flower; tuberose might for instance be a subgroup of Jasmine, while narcissus, lily of the valley, lily and hyacinth all belong to the Green series.

Carnation goes with Spices, mimosa with Anise, jonquil with Orange, gardenia with Fruit and so on. There is something to be said for a Bitter series (by analogy with gustation), typified by the crushed leaf of the bitter orange tree, But this note falls under three of the series we already have-Citrus, Orange and Green. Thus it scarcely needs a fourth one.

This classification appeals straightaway to the imagination and is easy to memorize. It allows odors to be pinpointed, arranged and objectively associated with one another. The practical value of this comes in the exchange of ideas, in placing a product one is talking about and in personifying an odor.

With this classification by reference categories one can cover the whole span of the existing palette. It becomes easy to arrange the majority of substances in current use in a way most people would agree with, and to fit in piecemeal new products as they appear on the market. To justify a new series there would have to be enough substances related to it and not to any existing series.

Each series contains the natural products inherent in or associated with it and the main chemical constituents of these essential oils. In the early stages one must avoid cramming the series. Otherwise the novice will lose his way among more strips than he can deal with or has time to examine in depth. I have no space to list here all my characteristic Series; so I will confine myself to the first-citrus, This can stand for the rest.

In any event there is no problem in finding in the literature the oils linked with our 15 series and their chemical constituents, limiting oneself to thirty or so names per series at the outside. The young perfumer has all his life to add to these.

This classification is an excellent teaching tool but valueless in the actual business of composition. The simple fact is one composes not with families but with individuals. The student must strive first to evaluate in practical tutorial work the sum total of the ways these individuals interact in mixtures, then to predict these interactions in research. This is the time when intuition comes to the fore. The combinations and permutations are so complex that there is no time to rationalize everything by checking it. Without intuition the intelligent technical study of composition be¬comes impossible, and research so hit and miss as to be sterile. Even more when it applies to an ar¬tistic creation. Since intuition is nurtured by experiment, the beginner must experiment relent¬lessly, studying each piece of work in depth to pull the lessons out of it and fix them in his mind. This in turn will spark off new intuitions In training creative perfumers, I use a set of exercises of increasing complexity for each series. Since the series themselves are arranged in order of increasing complexity, this training is pro¬ gressive in the fullest sense of the word, Neroli being the easiest of the first three series, these exercises start after initial studies of the third series,

Exercise No, 1 consists of a very simple mix of 6 products which are constituents of Neroli (the es¬sential oil distilled from orange blossom) but which I have carefully blended so as not to make things too easy. The student has to reproduce the composition by identifying-with his nose of course-the constituents and their proportions.

Exercise No, 2 might require him to match a mixture with the fragrance of orange blossom concrete, still with only a few ingredients.

Exercise No 3 is the reproduction of a mixture with the note of bergamot.

In the Rose series the exercises consist in matching increasingly complex rose fragrances, These may contain not only members of the Rose series but substances from the one he has already studied too, Thus, in moving from series to series, the exercises progressively make use of all substances already studied, But one must watch that the exercises are really progressive and that each is skillfully set up.

In this way the student's palette is broadened rapidly; the increasing demands made on his olfactory memory stretch it without overloading it, If, as they have to be, the exercises are well designed and graded, they will set the beginner a model he is apt to follow in his future work and will thus have a profound and lasting effect on his style.

With each exercise based on a formulation known to the tutor, there is no room for argument between master and pupil: the latter must constantly strive for a perfect match. This is no longer the case once the pupil goes on to study the great perfumes on the market.

Today, though, the availability of machines which break down formulas but having nothing to say about quality is making noses idle if not impotent. For the mind is apt to take the line of least resistance rather than raise its standards in step with its new aids.

The hard school of old is a thing of the past-whence the shortcomings of perfume composition today. Nonetheless a key rule that still holds good is to avoid prompting the student in his struggles to produce a match. Prompting stifles effort.

One must cunningly lead his student to find the missing substance or to trim the proportions on his own.

1		
Dip strips at 8.00	Lemon oil cold pressed	10%
	Bergamot oil cold pressed	10%
	Bergamot oil distilled	10%
	Orange oil cold pressed	10%
	Tangerine oil cold pressed	10%
2		
Dip strips at 09.00	Lemon oil cold pressed	10%
	Linden (lime) oil	10%
	Verbena (or lemongras) oil	10%
	Citral	1%
	Limonene	10%

3		
Dip strips at 10.00	Lemon oil cold pressed Linden (lime) oil Verbena (or lemongras) oil Citral Limonene	10% 10% 10% 1% 10%
4		
Dip strips at 11.00	Guinea orange oil Florida orange oil Bitter orange oil Decyl aldehyde Methyl anthranilate Limonene	10% 10% 10% 0.1% 1% 10%
5		
Dip strips at 12.00	Tangerine oil cold pressed Decyl aldehyde Methyl anthranilate N-methyl anthranilic acid methyl ester Limonene Dipthene	10% 0.1% 1% 1% 10% 10%

STUDY OF THE CITRUS SERIES

These substances are to be studied in 5 groups at 1 hour intervals, with a break of 5 to 10 minutes for fresh air between each group. The purposes of individual groups are these-

Group I-to compare the 4 oils lemon, bergamot. orange and tangerine.

Group II-to examine bergamot and its constituents.

Group III-to examine lemon, its analogs and their constituents.

Group IV-to study the oranges, their analogs and their constituents.

Group V-to study tangerine and its constituents.

Best to start work on each new series (or family of odors) on a Monday morning. After a good night's rest, breakfast at 06.00, pay the ritual visit and wash without perfume. Then do 30 minutes jogging with deep breathing exercises. When you get back, drink a glass of cold water and set to work,

All the strips needed will have been prepared the night before, with the names of substances and the date and time of dipping, The strips should be kept overnight in an odor-free box, Likewise sheets of paper will have been set out and ruled so that the impressions of successive smellings can be recorded as they arise, without losing a second. Generally speaking, follow the ground rules of 'Do we know just how the sense of smell works?', *The Young Perfumer and Scents,* Dragoco Report 4/62, pp. 93 to 103 This is very important.

Anybody who is about to smell unfamiliar substances for the first time must realize that these impressions are irreplaceable (more of this in a moment) and that calm concentration is the order of the day. During these studies of series, one simply must be incommunicado; lock the door and cut off the telephone, A ring at the critical moment, just as you are deep in concentration, about to hoist in a nuance and make a note of the sensation-and all is lost. That's that. You never *can* recapture the first fine careless rapture," Worse still, the irritation this untimely interruption arouses in you may well stop you regaining that indispensable composure and prevent you doing any further fruitful work on the study, which runs to a strict time schedule, A second lost may wreck a whole morning's work. Take this message on board and make sure everyone around you does too.

Now the smelling begins, Try to *pinpoint* and make a quick note of the *quality* and *character* of the odor (its note, its "shape", what it reminds you of or suggests to you); its *intensity,* its *diffusiveness* ("volume"); its *stability* or *instability;* the *evolution* of note and shape with time, This last must be followed through for several days or even weeks to determine *lingering quality,* All these *traits* constitute the *attributes* of the olfactory impression; they give it a *personality,* They *are inseparable* and must be seen as a *coherent entity*-but an entity which interacts in countless ways with the attributes of other odorants when introduced into a mixture.

Write down everything that occurs to you in the words which come naturally--even down-to earth-ones-if they help to build an *image,* to *define* and *fix* a thought, to *pin down the contours* of the odor *unambiguously.*

At all costs *steer clear of "roughly", "almost" and the like.* Seek out and find the words which define your sensory impression *unequivocally* so that twenty years later the same impression will bring the same words to mind. This is how-by *accurately* recording all olfactory memories and allowing the perfumer to retrieve them along with the name of the substance in question-the "odor card index" (or better still an electronic memory) becomes a really valuable tool. It is always easier to recall the smell than the name.

If the first smelling test does not suffice to pinpoint the sensory impression in words, try again at a different time and, if needs be, a different place. But never forget that *only the first impression* is *unsullied.*

So you must guard it, pay special attention to it and use a *lively mind* to record the first reactions for the sake of their *spontaneity,* even if deliberation afterwards leads you to correct them. It is in fact highly instructive to get to know the kind of errors of judgment you are liable to commit and to learn to mistrust yourself. This is the only way to develop a sound critical approach-and God knows you need that in our profession if you need it anywhere. But above all don't confuse spontaneity with impulsiveness.

Once the first impression is over, it may occasionally be possible to recapture a further unsullied impression-a very long time after, once the memory has completely gone, or indeed if lack of warning results in surprise.

It is this surprise effect that one must strive to bring about in studying compositions. He can space the tests out or stop doing them for several months now and then. This allows one to break the bounds of *obsession*, to forget shape and formula, to try and come back to the abandoned experiment with a *fresh and impartial judgment*. This is difficult, it calls for strength of character, doggedness and hard intellectual integrity. But what a fine schooling! What an asset and a comfort it is to feel oneself veering back towards *objectivity and certainty of judgment*. Breaking of a study does not mean a life of idleness. A change is as good as a rest.

The "nose" has to watch carefully how he behaves and how his olfactory membrane is standing up to the demands he makes on it. Never hold strips less than half an inch from the end of your nostrils; otherwise you will contaminate your nose and impair subsequent smellings.

NEVER DO YOUR SMELLING IN A LABORATORY OR ANYWHERE WITH A SMELL

I wonder how many houses give their perfumers the chance to observe this most basic of ground rules. One has to lock himself in bitter combat with the insidious assault of *smells-whatever they may be.*

Any outside odor clouds and colors judgment. So smoking goes on the list straightaway. Never smell too soon after eating or drinking. After every meal, brush your teeth carefully to re move every trace of food which is liable to decompose and thus to smell, rinse you r mouth out with a pinch of bicarbonate of soda in a glass of lukewarm water to neutralize any acidity there may be. Use nonperfumed toothpaste. Don't start the smelling session for at least an hour after eating; otherwise your stomach will still be giving off smells of food.

These constraints have to be taken account of in laying down schedules for smelling tests, so that no time is wasted after a meal if there is smelling to be done. But do your really serious smelling in the morning.

The substances in each series must be followed through on the strip as long as any perceptible odor remains. Sniff dried out strips in the morning, after washing and taking the air, but before breakfast. This work has to be done at home. Naturally one starts with the driest and weakest-smelling strips.

During examination of the series of odors, reoxygenation sessions between each group are essential to ensure adequate *regeneration* of the olfactory mucous membrane and to restore mental relaxation, Best to use part of the time for a short jog and the rest, on the way back to the office, for "walking down" with the aid of deep breathing exercises. One also needs to know how to regenerate the membrane during examination of the groups, between two phases of smelling, by going over to the window and taking great gulps of fresh air These details are very important if the procedure is to be effective and reliable.

Attention! The smellings, to be accurate, should *be short*. You should smell by series of 2 to 3 inhalings each of 2 to 4 seconds, as sensibility is rapidly exhausted. If you are not able to limit the time of each smelling and their *number,* there is no guarantee for an accurate judgment with all its negative consequences.

In most of the series there are substances varying in concentration from 10% to 0.1 %. One must never lose sight of this, bearing it in mind throughout the examination and revising his notes on the odors in the light of it. A 10% concentration should be regarded as *a standard dilution,* applicable for the great majority of substances. It has the advantage of being dilute enough to allow the "shape" of the odorant to express Itself. With most substances it runs no risk of insulting or overloading the mucous membrane in a way that would affect *judgment*. By contrast this concentration is not too far from that of perfumes, so that each substance can be evaluated under evaporating conditions comparable to those of a perfume. This is extremely important in making a valid judgment.

Thus the *image* of each product ultimately to be engraved in the perfumer's memory will be *the image 01 the substance at standard dilution-10 %, 1 %, 0.1 %* or whatever, depending on the odor intensity of the substance. This is of key importance for convenience, consistency and effectiveness in evaluating, identifying and classifying---in a word for the whole span of mental operations.

This technique likewise gives the best results when the problem is to *evaluate* substances with respect to others When the perfumer thinks of linalool, he will think of linalool at 10%; when he thinks of methyl anthranilate, he will call up the image of the substance at 1 %, of its intensity at this concentration, and its shape. He will thus get used to *estimating,* for instance, how much methyl anthranilate at 1 % can be tolerated by 1000 g of linalool at 10%. From there he will go on to estimate what proportion of other substances he might successively introduce into this initial mixture-but always thinking in terms of the standard dilution of each substance.

In this way the danger of misappreciating the optimum proportions for marrying up odorants will be far less than if one bases his thinking on substances at 100 %, a form in which they cannot be compared. One can estimate the relative intensity of decylaldehyde at 0.1 %, maybe even at 1 %; but he cannot do this by smelling the substance at 100 %, for its excess intensity would block the olfactory system, protected as it is by an inhibiting mechanism. This would prevent him making any valid appreciation of the substance's actual strength, its practical properties or its shape.

Drawing up an empirical table which gives only the "equilibrium" proportions of binary or at most ternary mixtures is a simplistic solution fraught with danger. Simplistic because the interactions which arise in a medium of fifty or a hundred constituents like a perfume recipe have nothing whatever in common with the interactions between just two, three or four substances. So it is no good knowing that A and B are balanced; introducing the rest of the alphabet puts the whole thing back into the melting pot.

This technique has even less to be said for it to the extent that the build up of a perfume is not a matter of balance. Composing a formulation is not about "balancing" the various constituents but about *marrying* them harmoniously in such a way that they combine to create an *olfactory shape* with such and such a set of *characteristics*. No matter whether the constituents are "balanced" or not-the more so since one can put dissonances to good use. The thing to look for is not so much the balance of the constituents but *the way they act in concert* and the effect of the chords they produce-the melodic aspect of the composition. A composition which is nothing more than balanced adds up to a perfume which is static, inhibited and deadpan. A well composed perfume must *live* and *move*.

Tedious as it may be to draw up, a table like this is a dangerous approach for another reason too-It tends to make the user idle minded. The way things are, a perfumer who relies entirely on this table and combines his constituents in the proportions shown in it would be in for a nasty shock. Just adding a second pair of substances to the initial "balanced" binary mixture is enough to topple the balance; for that was established between two products, not between four The more ready balanced pairs one adds, the greater the chaos will be.

But the worst thing of all is that a perfumer who used this approach would be turning away from the path of studying *each of his substances* in its *relationships* with *all* the others. So he would not be in a position to fix these relationships in his memory in a way that would build up little by little. experiment by experiment, an *image* for each product which at once comprehends its *outward appearance, its scope for relationships,* in fact its *whole personality.* It is the whole personality that the perfumer is going to *play about with.* So it is the whole personality which he has to *assimilate and integrate into his own being* to the point where he is so *familiar* with it that he can readily *juggle* with it.

Once he knows each substance as well as this he will be able, when on the point of introducing it into a new combination, *to predict by intuition* whether or not it will be a success and, if so, in what proportion. Intuition is not a miracle but a *spark* which will only be discharged when one has built up a large enough charge of *knowledge, experience, deliberations and meditation.* And it is only by these flashes of intuition that the perfumer *moves* forward, making his way slowly but surely along the path of a study.

Some series of odors contain crystalline substances. Even less than in the case of liquids can the *action* (olfactory shape, strength, behavior) of these crystals be evaluated in the 100 % state. Their vapor pressure is as a rule lower than that of liquids and the development of their odor with time cannot be followed in the crystalline form. When examining them, and afterwards when recording them. one must not forget to associate their *crystalline form with their image,* for this form in its own right has a part to play in a composition. Crystalline substances often *have* a *very* high boiling point and evaporate *very* slowly. Thus they exercise a *braking* effect on the other constituents of the formulation quite apart from their *lingering properties* as *such,* which also exercise a *delaying effect.*

But one must not generalize about the behavior of crystalline substances. The important thing is that their image should reflect their *specific* effect with all its *quirks.* Thus heliotropin melts at 37°C and boils at 263°C. Coumarin melts at 68°C and boils at 291°C, but its lingering power is much greater than that of heliotropin-far more so than the difference in boiling point would lead one to suppose. The lingering properties of vanillin which melts at 83°C and boils at 285°C are still greater, in fact among the highest to be found in aroma chemicals, Yet in a fragrance its influence will make itself felt very markedly the moment evaporation begins-just another of nature's tricks.

Menthol melts at 51°C and boils at 211°C, while camphor melts at 179°C and boils at 208°C, Despite their crystalline form, these substances have *intense* odors, specially camphor although it only melts at 179°C. A further *peculiarity* of camphor is the proximity of its boiling and melting points, the transience of its liquid phase and its ability to *sublimate* at room temperature, Crystals they may be, but these two substances, menthol and camphor, cannot be seen as having any particularly delaying effect on a fragrance; they tend rather to provide it with *impulse.*

The three nitro musks-ambrette (MP 83°C), ketone (MP137°C) and xylene (MP 114°C)-have extreme lingering properties, notably musk ketone, Certainly they have some delaying effect. But they tend rather to *detach* themselves excessively towards the end of evaporation and it is mainly them that one smells; broadly speaking they do not seem to do much to conserve the shape of the fragrance as a whole.

In the Violet series we come to the ionones and methyl ionones which have to be smelled under precautions. These substances have the property of rapidly inhibiting the olfactory mucous membrane, so it is advisable not to linger for more than one second over each strip impregnated with them. Otherwise one may smell them wrong or not at all. Since they are among the noblest series in perfume chemistry, it would be a pity not to be in a position to evaluate them properly and thus to use them well,

Within the same Series, the various nonadienes likewise call for great precautions, They are olfactory dynamite. Only extreme dilution to 0.1 % or even 0.01 % allows one to appreciate their excellence,

PROPOS & PARFUMS

By Edmond Roudnitska, Perfume Creator

W.W.W - Dominique Viot - 27/06/96

DEAR FRIENDS, I address all of you who blend, make, direct and decide in the field of perfumery products. For you are all part of an indissociable whole, together we are perfume, and you all depend on one another.

In the interests of everyone, and of the industry as a whole, you need to work closely together, to combine your efforts rather than to fight, so that the perfumery trade can benefit from your varied and unique contributions.

So before thinking in terms of profit, glory and power, think of perfume. If you subordinate everything else to what is - let's not forget it - the raison d'être of our profession, perfumery will go so well that the other requisites will follow as bonuses, whitout wounding anyone's honour or pride.

Do not restrict yourselves simply to your own particular task. Broaden your horizons, initially so as to understand what the person next to you does, so that you can unite with that person in producing the finished, shared product. Then broaden your view to take in the whole universe, to those things wich you can only benefit from if you return to the deeps roots of authentic values, those which inspire all great works, both in the past and the future.

True culture, which is far more than mere erudition, is the only foundation on which anything solid can ever be built. The long-lasting nature of the works that result from it is the only value which never disappoints. It is because man is mortal, and because he is man, that he aspires to immortality. So be men ; work for the temporal or the ephemeral. Think about leaving something behind you, so you can depart with the conforting feeling of having been useful. Vanity can only leave on embittered, because it is never satisfied.

EDMOND ROUDNITSKA : A LEGACY

Mr. Roudnitska started his career in 1926. Since that time many changes have taken place surrounding the point of view of art, creativity and perfumery. His interests were broad and encompassed collaborating with researchers in olfactory studies, ecological restoration and the arts.

Much of his life was spent defending the creative side of the perfumer's artistic copyright, thus giving legal legitimacy to this "art of perfumery." In his book "Le Parfum" (The Perfume) he comments on the judgement rendered by the Paris appeal court. In "Le Droit du Parfum" (The Right of Perfume) he expands on this topic.

On of his greatest contributions to the art of perfumery was the concept of shapes and form in fragrance creation that he began to discuss in 1944. This concept has been widely recognized throughout the fragrance industry. He told me, "what counts is that you think of the form and use the minimum amount of products in the creation of a perfume." In "L'Esthétique en Question" (The Question of Esthetics), he emphasized many times that the survival of perfumery as an art depended upon an educated public.

The shape of a perfume derives from an aesthetic combination chosen and desired by the perfumer. The fragrance formulation results from an assembly of constituents which gives rise to an overall phenomenon which is anything but simple addition.

Everything Edmond Roudnitska wrote about perfume was designed to develop our awareness of the world within; our inner space, our feeling about ourselves, our desires and ambitions. How perfume can help us deal with these aspects of our being is its most important function. If a perfume is well made with a clear and distinct form, it can help us gain awareness and mental clarity. Roudnitska speaks always of perfume not "fragrance" which lacks the profound significance and emotional impact of a real perfume - a work of art, a masterpiece.

Much of his later writings reiterate the need for the public's education and ability to understand and recognize the beautiful. "The right of everyone to live in beauty is affirmed and one can see the public's appetite for beautiful things growing."

For the perfumer with a well-trained consciousness the sense of a smell is above all a sense of knowledge, of verification, of entry into a special universe infinitely rich in signs which the composer turns to his aesthetic ends.

Mr. Roudnitska was a master perfumer who created "Femme", "Chiffon" and "Moustache" for Marcel Rochas in 1943, 1946 and 1948. For Christian Dior he created "Eau Fraîche", "Diorissimo", "Eau Sauvage" and "Diorella", among others.

He welcomed visitors, many were aspiring perfumers. He once said "the contact is immediate with the young and there is instant rapport which comes from the earth and soul and that is the level that I am working from". His greatest influence was that he was one of the few independent perfumers and he created for his own pleasure, having complete freedom to make diverse and unusual perfumes without regard to price or mass market acceptance.

We are grateful to have had the opportunity to meet with such an incredible man and shall miss him as we know many others throughout the world shall as well. To sum up his legacy in these few words is impossible. He persevered against all odds in pursuit of his dream.

In fond memory, Christine Malcolm & Kathryns Degraff

AESTHETIC CHARACTERS OF THE FRAGRANCE

© Edmond ROUDNITSKA

What is the criterion of the artwork? Concurrent relations, that is, the existence of proportions between the parts of a whole and of each part at all? If anyone still doubts that this is the definition of a great perfume is that we have been very eloquent. In order to leave nothing in the shadows, let us see with Etienne Souriau on what conditions an art can be recognized. There are five, he assures us:

First condition

"The exciting attack of a special psychophysiological sensitivity (sense of colors for painting, movement for dance, sounds for music, etc.), sufficiently developed to interest, move, excite the imagination or thought, provoking aesthetic pleasures "

(The music of perfumes, conference, Paris, May 1951, Perfumery Industry, February 1952)

The sense of smell is a sensitivity that can not be more special because, being a chemical sense, its reactions are finer than those of a mechanical or physical sense. We now know the exceptional capacity for discrimination that gives this instrument of measurement and estimation if not an absolute precision at least a precious subtlety to play as abundant palette as ours.

That is why Etienne Souriau recognizes that the olfactory sense is extremely delicate, that it is susceptible to be excited by infinitesimal quantities of matter, to give clearly aesthetic impressions and to perfect oneself in this direction. Properly trained, he says, the sense of smell does not yield to hearing as it is applied in musical hearing, and its sensorial data possess affective properties on the occasion of prodigious.

Etienne Souriau shows a lot of insight when he considers that the olfactory universe is infinitely varied, delicate and nuanced on the purely sensory level but also rich in affective powers, and that this universe is more, in the intellectuality and particularly on the imaginative plan, endowed with strong evocative and representative values. On this first point he concludes: "Do not hesitate to recognize that the sense of smell is an aesthetic sense in the highest degree and that it presents as much and better than any other sense, all that can serve to support an art full"

(The music of perfumes)

We will doubt all the less that what determines the work, as we have already shown, is neither the meaning nor the material but what a thought, playing with forms, will make of the second to the help from the first. The work of art is not conceived by meaning but by the spirit of man. This is to be understood once and for all that an artist composer is not a "nose", as it has been written so many times, but a brain interested in olfactory forms and who treats with taste, which is no longer the same thing.

Second condition

"Possibility for the various qualitative data of this sense, to order, to manifest relationships, to lend itself to architectural arrangements, in short to provide more or less a range or a palette: a limited system, usually scalar, of quality. likely to be in a certain order assembled, to combine in multiple variations always more or less well organized "

(The music of perfumes)

This condition is fulfilled when the composer establishes a formula which, without leaving anything to chance, is based on a deliberate and ordered choice of components which he determines with a minimum of trial and error proportions to make a harmonious and meaningful whole, it is to say having a characteristic, original and pleasant form. Pleasant meaning who likes and not "nice", as it looks like a cabbage cream. But note that in an olfactory composition each component is not significant by itself, it becomes only by its relationship with others and especially with the whole. The olfactory arrangements are thus based on the play of the associated odorous qualities, conjugated, they express themselves by reports notified in a formula where the proportion of each component appears. It is the spirit that has presided over the creation of the few great perfumes of the century, scarcely more than a dozen so far, and it is remarkable that universal opinion has seldom been deceived, which has been able to distinguish it.

Our materials give rise separately to singular olfactory forms which are inscribed in our memory, like sounds and timbres in that of the musician. By thinking we evoke these forms and we imagine to combine them, to combine them, to build with them olfactory sets that will give rise to a global phenomenon, the perfume, which will be very different from a simple addition of its elements.

To compose we do not need to feel our materials concretely, sensorially, the mental effort is enough to evoke them and the combination of the forms they represent can be done as well as in the abstract, without possibility of control immediate material. At this stage our effort of abstraction is comparable to that of the music composer, who also thinks his forms pencil in his hand. The musician then still has an advantage over us, is that he can immediately check his chords by humming or strumming, while we can do it only after a laborious weighing (two to three hours) components listed in the formula, a subsequent alcoholic solution of the concentrate thus obtained and a sufficient maceration thereof. Which represents in all several weeks of waiting. This is why our mental effort is more like that of the musician composing a symphony, and which must then embrace in a single abstract vision no longer simple local agreements quickly verifiable but large composite sets whose verification will be laborious.

On the other hand, to express the value of the attributes of the sounds: pitch, intensity, duration, the musician has a particular sign for each of them, he can adjust them precisely, just as he chooses practically without constraints. his tempo, that is, his speed of execution. With a single number in front of each component, we must take into account its quality, intensity, duration, affinities, and we suffer its volatility which we must estimate the influence it will have on other volatilities and the one she herself will suffer from them. A single report to translate all these factors while we must necessarily have each smell a multipartite representation. We mentally brew hundreds of smells and each time we introduce one into a formula it is under its multiple aspect that we must consider its behavior with respect to the other components and the role it will play as a whole. . The perfume is therefore a construction of the mind.

That our materials can be ordered, we have just shown it and it depends only on our intelligence to provoke it. Complex as it is, defined body or natural essence, each raw material has for each composer a specific physiognomy which he classifies in a system of values which is necessarily personal to him (But is it not so in all During each study, he mentally extracts from this system the elements of formula which seem to him to be appropriate to the roles to be fulfilled or to the desired effects, and thus from studies in study, thanks to the observations and reflections they have aroused, that the composer modifies, completes, perfects, consolidates, justifies his system of values.

What is the need for the system to be limited? Those of the sculptor, the architect, the artists-decorators, and even the painter are they really? The musician himself plans to expand his. The "more or less" a range or palette, Etienne Souriau, shows that few materials lend themselves to it; some are even more reluctant than ours in this regard. Most signs do not really have their grouping and order to the personal vision and the will of the artist who manipulates them. Except specialists (television for example) who treats them in computer, the painters do not work with wavelengths but with an infinity of nuances very superior to the seven colors of the spectrum and that they handle with their only intuition.

Etienne Souriau also admits variations "more or less" well organized because we must make out the vagueness and confusion in activities like those we just mentioned. This tolerance of the criterion which characterizes the work of art is essential in all the arts, if only according to the hierarchy which is established naturally between the works and between the authors. There is no doubt that the assemblage of signs will be less likely to be well organized on the canvas of a Sunday painter than on that of a Cezanne. Between the two will be spread out an infinity of values to many of which one will not however deny the quality of work of art.

The same goes for perfume. Formerly the small number of French perfumes put on the market was the fact of composers of great talent, both authors and owners. It was a work of art; when it was not, their career was brief because the taste of the public was then severe and enlightened. Today the destinies of Perfumery have changed hands; if there are still some composers, more potential than active, there are no longer any owners. All that remains is that businessmen, inexperienced and influenced by imported and inadequate commercial methods, blindly and recklessly leave the multitudes of perfumes elaborated - in pharmacies specialized in industrial perfumes - by the mercenaries of whom we have spoken. So there are still some works of art, rare, that have been able to cross the trade barriers, and there is ..., all the rest, supported and promoted by means that the artistic conscience reproves but in front of which the taste of the public, disoriented by numbers, eventually succumbed.

It is sad for the art of perfume, and for all those who love it but it is sadly not less in the field of painting and sculpture where we see commercialism support and promote the imposture exploiting the snobbery of fools or the ignorance of the innocent. The saving purification can come only from the revolt of a more informed and uncompromising public. But with what concerns us, with a greater generalization of chemically defined bodies in the structure of perfumes, and provided that they are composed by artists, we will move towards a better controllable technique, or the "blur" and the "confused" will have less and less reason to be.

Third condition

"The existence of technical means, including instrumental procedures of execution and presentation, allowing the artist to achieve with a certain precision and a certain certainty, and to present to the public with sufficient

vivacity in the delicacy, the the intensity or significance of the reports, either a unique original work or a repeatable work according to a constitutive scenario, which is itself original and unique. "

We have throughout this book and again on the previous pages, multiplied the opportunities to familiarize the reader with our technique of composition; we even tried to involve him in our creative concerns. We have seen the composer conceive of his perfume in the abstract, translate his thought by a formula that enumerates products and their proportions, products whose aesthetic combination (and not the simple physical mixture) must realize the olfactory form conceived by the composer.

All the arts do not require instrumental procedures of performance and presentation. Even the musician, as we have seen, can be content with paper and pencil to compose. To achieve with precision and certainty our artistic designs, it took us tens of years of effort and sacrifices of all kinds. We have established that it was possible but we must agree that the example seems difficult to follow in our time and in our profession because the emulators do not abound. It is true that every thought, to the extent that it is an example, appears as an aggression against established ideas.

(José Lemaire, Cyber File, p.69.)

At his conference twenty-five years ago, Professor Souriau had begun to apply to the perfumery methods of composition, inspired by those used in musical composition. But he had limited himself to the melodic comparison, very utopian in the conditions in which he envisaged it, whereas a contrapuntal image would have been more valid because it is in fact what happens in our perfumes where very melodies slow are superimposed during the main scene. But at the time, Etienne Souriau could not suspect what the olfactory composition, on which we stopped informing since.

The vivacity, the delicacy, the intensity and the signifiance of the relations which characterize our great perfumes are so notorious that one would have to discuss commonplaces. We will spare him from the reader, all the foregoing having thoroughly familiarized him with these considerations.

The formula of a great perfume is obviously a unique original work, which, at the application becomes the constitutive scenario that will allow to make the perfume and to repeat indefinitely this work, just as do the score of the musician or the manuscript of the 'writer. We must protect this formula against plagiarists.

The practical realization is to weigh the constituents of the formula in the proportions chosen by the composer, to mix these products and to dilute a few weeks later this mixture (called concentrated) with alcohol, in a proportion very carefully chosen too because this proportion will affect the final shape of the perfume.

Before checking and presenting the composer should in principle (for reasons similar to those which condition the great wines) let the alcoholic solution age for a time varying from three to six months depending on the structure of the composition. in order to judge it, it will then have to evaporate it on a suitable support, studying at length and attentively all the stages of evaporation of the composition; which represents several weeks of serious control and very critical examination.

(Ceramists also suffer long delays before they can judge their works.)

To present his perfume it will be enough for the composer to vaporize objects or people; but the amateur must also, if he wants to make a valid judgment, observe and carefully follow the composition as the perfumer does. Too many people have the lazy habit of limiting themselves to a very superficial examination, which involves putting one's nose under the neck of the bottle or putting a drop on the back of the hand and rubbing the skin, which may cause Amino reactions and to emit a foul odor, or at least to denature the scent. Our compositions are delicate mixtures that are not made to undergo such treatment. A classy perfume should be used with the vaporizer; to be satisfied with examining at the bottleneck a deceptive outlet is to expose oneself to contemplating a facade which masks nothingness, or: to deprive oneself of discovering the wonders that follow and will last.

Another serious error of presentation consists, as happens in all the perfume shops, to make feel in quick succession, without rest very long intermediaries, several perfumes that have not been designed at all to oppose one another. others but on the contrary to be felt in isolation. The laws of the relativity of sensations pitilessly sanction this error. Hearing is much less likely since during a concert the most different pieces follow each other without our ear suffering and without it being very difficult to establish the program so that the pieces either do not suffer. . Nevertheless it would not come to a sensible person the idea to listen simultaneously two different symphonies.

Fourth condition

"Historical and social conditions have sufficiently favored the development of this sensitivity and these techniques so that the education of creators, performers and the public is adequate to high aesthetic requirements."

The use of perfumes since time immemorial, the advent in the twentieth century of a modern perfumery, and its prodigious development in the world under the impulse of the French perfumers, should have led to a correlative development of the sensitivity and the imagination for both creators and users. It has not been exactly so for reasons that we have amply developed in L'Intimité du Parfum and on which it is useless to return.

But nothing, absolutely nothing, except the apathy of the main interested parties, does not prevent the education of one or the other being realized. Because young potential creators are still full of enthusiasm, and the immense curiosity of the public towards us is a sure guarantee of the interest that it would bring to such an effort.

What must be persuaded is that the formation of the creator is long, that the education of the public, for which we have been fighting for thirty years, must be patient, so let's waste no time and say that 1974, with L ' Intimacy of Perfume, will have been the year I of education since for the first time a qualified perfumer addressed himself directly to the public.

But despite the absence of any education, any real information, despite even and especially, the sum of all the nonsense that has been spread around the world on perfume, how not to be confused by the spirit of penetration, by the intuition we can say, of which we show with respect to the great authentic perfumes and that for six or seven decades, all the public of the world and singularly that of France.

65

These historical conditions are a reality that can not be denied. Because if the Sixth and the Ninth, without any other precision on display, fill the halls anywhere in the world, we can not refuse the same universality to the great perfumes of which millions of bottles are distributed in the most important corners. of the universe. Putting this election on the sole account of sensuality (which some estheticians also annex) would be as blind as attributing the universal infatuation for the Sixth Symphony solely to affectivity, or to a unanimous and superior intellectual comprehension that the audiences Beethoven's folk can not explain.

When, in such diverse climates, in spite of traditions, habits of life and food traditions, so different, there is a certain unanimity in the field of taste, it is the incontestable proof that the object which benefits from it is a matter of beauty. , of the universal beautiful. A fortiori aesthetics.

Fifth condition

"Men of talent or genius having found in these techniques and in this kind work an attractive and important means of expression (which, of course, is only possible when the embryonic period of research, trial and error and drafts is clearly outdated). "

These men, few since the beginning of the century since they count on the fingers of one hand, have existed in our profession, in France exclusively, and their creations, since, esséiment around the world where they contribute to train the image of France. There is, there, I answer, not the shadow of chauvinism but the statement of a patent fact. Is this negligible? And for negligible the archetypes that French perfumers have thus procured to lesser ones, to those of successive generations, to their foreign confreres? All these seeds to be rare, are they less precious, less generous, their diminished germinative power, their value as a questionable example? On the contrary, they are irreplaceable milestones on the road to art and human thought.

The end of the embryonic period of Perfumery is at the very beginning of the twentieth century. The Oregano, of Coty, in 1905, his Cyprus, in 1917, were no longer drafts but masterpieces of harmony. Oregano, for example, in addition to the natural essences and chemical substances already contained in its formula, was a very heterogeneous assemblage of complex bases, that is to say mixtures prepared by the various suppliers of Coty. He was totally ignorant of the composition of these bases, so he could not know a priori how their constituents would behave between them, on the one hand, and with regard to the other constituents he had chosen, known to him there, on the other hand. It took a serious dose of intuition and genius for his final composition, the Oregano, to be a model of unity and a beautiful, very distinctive, original form that perfumed the entire world for over twenty years. years, then received a universal assent.

The same can be said of Arpège created in 1925 and especially Scandal and Rumeur who were on the eve of the last war. These last two were literally "tawny" perfumes. And doubly because if the boldness and the realism of their composition made them perfectly assimilate to the painting of "beasts", the exceptional harmony that their author had also had the genius and the taste to create around the leather note, made call to animal odor components, extremely ungrateful use and which were well, they, odors of big cats.

The mastery implied by such formulas could only be the result of an accomplished artist, who would prove not only his talent but above all that the olfactory composition can be a remarkable means of expression. If some judges of perfume are reluctant to understand (why?), The public, on the occasion cited, showed once again by his clairvoyant and enthusiastic welcome, what sensitivity and what daring he is also able in presence of authentically beautiful works. And the proof that the public is fully aware is that it abandoned these masterpieces as soon as their quality and their form deteriorated, probably because their creator was no longer there to defend their integrity. .

The composition of perfumes is abstract art par excellence, hence its difficult intelligence, but it is also par excellence the art of happy proportions, hence the extreme intellectual enjoyment it provides to all those who know thus to appreciate it but how much also to him who learned to practice it.

THE WORLD OF FRENCH PERFUME.

By Edmond Roudnitska

HOW DID WE GET WHERE WE ARE? WHERE DO WE GO NEXT?

Introduction

Early this century French perfumers were three men rolled into one - composers of perfume. managers and salesmen. In other words, they were at once connoisseurs and businessmen. If they created their perfumes mainly to please the Parisian clientele, this was because that clientele too consisted first and foremost of connoisseurs. It was a pi lot market par excellence, for its decision to accept or reject set the tone for the whole world.

If French perfumes not long-ago occupied pride of place on the world market, it was because they were first-rate perfumes-the best by far. If they appear to be losing their monopoly today, it is because of the lack of originality of too many new launches, rather than because of improvement in the quality of foreign perfumes. It is to the richness of our own rich inheritance that we must look for inspiration -nowhere else.

Nor is it, as is sometimes said, the foreign clientele which has become much more critical. It is rather the Parisian clientele which has fewer tine models to attune its taste to than it had in the old days. Unless we outdo our foreign competitors in quality, we will find it hard to stand up to their striking power.

Does democratizing perfume make it trite?

Perfume is accused of being a "luxury," reserved for an elite, instead of being accessible to all.

Does the fact that the less well-off cannot afford it makes selling caviar a crime? (I hate caviar, by the way.) Would it be right for every corner shop to stock it? They could, of course. But who would they sell it to? Caviar is expensive for good economic reasons -notably that sturgeons have obstinately rejected Stakhanovite principles.

Or again, are we to ban mahogany, rosewood, ebony, and burr walnut furniture because it is beyond the reach of ordinary people? These woods are dear because they are something special. Are we to insist that all furniture must be made of softwood and thus penalize the third world which supplies us with 1hese precious woods?

In art galleries, do we find that all works of the same size cost the same? Do we have to be shocked because a work is priced on its merit and its creator's talent? Are we then to denounce talent? The absurdity of these questions of mine stands up like a sore thumb.

Perfumes are likewise works of art which may have greater or less merit, more or less success, and the creation of which has call-ed for more or less time and effort. The cost price of their raw materials - excluding talent and overheads-may vary with their com-position by as much as a factor of 50. Tell me, please, the moral principle which impels us to discourage a creator of perfumes from working with expensive materials if he considers them essential to the embodiment of his idea, and if, as a result, he finds a backer capable of selling his creation. Nor, surely, does any aesthetic principle drive us to claim that the cost price of all perfumes should be leveled.

The expression "luxury perfume" is as lacking in real meaning as luxury art, luxury music, luxury literature and the like would be. Coining and spreading it in the way peo-ple did was a regrettable error. Some per-fume formulations come very dear, others very cheap. But this does not automatically make the first ones better than the second; what does that is the competence and talent of the perfumer who composed them. I have often said that, given talent, one can com-pose good perfumes at low prices. But this does not mean, please, that all cheap per-fumes are necessarily good ones.

If the cost price of perfumes varies and the selling price follows this variation, there are diverse types of customer too, with differing tastes and different sizes of pocket. To meet this span of needs, it is thus logical to com-pose perfumes at price levels which match the various cost prices and the diverse groups of customers. There is no room at all here for some kind of egalitarian mythos. After all, one does not find any such thing in a country like the Soviet Union, which in fact has long had one of the broadest salary ranges in the world. In parallel with local pro-duction, the Soviets import very large quan-tities of the great French perfumes for the simple reason that the privileged are in a position to buy them.

The French perfume world has always been marked by a certain degree of democratization, with brands like Pivet, Millot (hairdress-ing products), Roger et Gallet, Lubin, Cheramy, Bourjois, and Coryse Salome-and nowadays Avon, Yves Rocher and the like. The same is true abroad. The grandes marques too bring out some items within reach of the average pocket. Democratiza-tion is thus a nonexistent problem- or at most one of the past. The fuss about it is even less justified inasmuch as 70 to 80% of the great French perfumes are sold abroad, bringing substantial amounts of foreign currency into our country's economy.

Since a great perfume often costs a lot to produce, it cannot be sold for a song. What the consumer needs to know is that, of the price the retailer charges him for a flask of perfume, the State takes over 30%, and the retailer himself keeps at least 40c. The balance less than 30°10 belt noted, is all that Is left to the producer to cover amortization of fixed assets, research, purchase of raw materials, manufacture, presentation, administration, advertising and profit This amounts to saying thal those on whom lhe creation and production of great perfumes depend spent their lives walking a tightrope. They are not the ones to complain to about the price of perfumes.

One might add that lhe mass-market inter-national brands lose little time in offering more or less spoiled copies of the great per-fumes on the cheap. This practice is neither moral nor legal: but it adds up to another form of democratization.

It is no use getting angry at seeing an expensive article sold at a high price to a clientele which may be relatively limited in numbers yet remains very substantial, both in France and in the world as a whole. The country's economic interests, likewise its cultural image - in which perfume itself plays a part, being indeed no mean ambassador-call for a defence of the true image of this product. There is no need to use the words "image" or "prestige" in a deprecatory way; these terms are simply vehicles for sales, in par-ticular for export sales.

This is why quality products accelerate aimed at a particular market must be presented and offered for sale in matching surroundings. And if the hypermarkets claim they can sell these products readily to their customers. this simply means that such perfumes are 101 in fact as far beyond the average pocket as they are made out to be.

By contrast to democratize is not to debase. It is to make beautiful things accessible without degrading them -and this calls for even more talent, stretching the creative perfumer's art to the full. We sometimes confuse potency with quality; the public, though, does not in the main care tor violence and gets out of 1he way as fast as it can. The fact is that it is easier to make a per-fume which "shrieks" than one which is lovely and pleasing.

The best example of successful democratization is that of sung music by the cinema. The filming of Jean-Pierre Ponnelle's 1976 production of The Marriage of Figaro put this opera within the reach of the public at large via television. So far from suffering from this extension of form, the work was enhanced by the Vienna Philharmonic, by a brilliant cast, and by settings far more varied and natural than those of a stage set. One thus had movements in space where more lifelike characters could act more freely in expressing the whole of their emo-tion in a quite natural way. This revealed the full beauty of the piece. Then there was the camera work- mobility, shifts, choice of close-ups and the like- together with the art of editing, in harmonizing libretto, voices, images, and music in a larger- than-life way to present a total and sublime operatic spectacle.

The vast audiences ensured by the small screen-and now the large one-easily covered the costs of an exceptional production which would have ruined any opera house. Since then, Don Giovanni, La Traviata, and Carmen have conquered the public everywhere, providing still further demonstration of what real, intelligent democratization means.

The "dream" element we attribute to perfume is an easy way out.

True, smells can readily awaken memories, that the perfume of a woman loved and lost can revive the ardors of a man of sensibility. But women today are more positive. What they are looking for in a perfume is not so much a dream as a reality, an intimate complement to their beauty and their adornment, a sensation in-dividual to themselves to begin with, then to be shared with those around them. With her perfume as with her dress and her hair-do, a woman seeks first to please herself, before she worries about pleasing others.

Only if she is pleasing to herself will she feel herself able to please others. And sensuous thoughts are only one element in this desire. She will be at least as interested in pleasing her woman friends, to bowling them over, and in making her rivals jealous, as in at-tracting men.

Again, as in her dress, a woman sees in her perfume a pretty shape which suits her, which highlights her individuality, which makes a personality of her, which gets her recognized. And even more than the dress, the perfume forms an intimate part of her. All that is nicely calculated- no way a dream, but self-affirmation. A song can make one dream too; but this does not alter the fact that its true reality Iies in the poesy of its lyric, the accompanying music, and of course the vibrato of the voice that sings it.

In sum, those who buy perfume come from every walk of life. As has long been the case, there must be perfumes for all tastes and for all pockets. The down-market perfumes do not sell any less of a "dream" than the great ones - rather more, in fact, for a great per-fume's aim in life leans further towards the aesthetic than critics and media men, stuffed with ideas which are at once received and false, often give them credit for. The sensual and sentimental considerations I discussed a little earlier are just a "plus value," added to the aesthetic construction of which many women are perfectly well aware, at intellectual and intuitive levels alike. Their perception of perfume is not too different from that of the creative perfumer when he stands back and takes a long cool look at his work. This is what enables him to understand his clients.

Have women really become less faithful to their perfumes?

To the extent that this phenomenon of readiness to change perfumes has become widespread, it may well be an import from America. There a perfume seemingly tends to become identified with a fashion and thus to enjoy a shorter-lived success than in Europe. But, in France as in America, the real reasons for this pattern are not those generally put forward. It does not stem either from the evolution of tastes, or from changes in popular mores. In fact, it is nothing to do with the public. Let me begin by pointing out that the alleged readiness for change of the American public does not stop them continuing to give a very warm welcome to perfumes like Emeraude (1921), No 5 (1923), Shalimar (1925), or L'Air du Temps (1947). The American market takes more of these than the French does. Most of the so-called novelties are nothing more than mixtures without any personality to speak of, or cheap copies.

In France as in the United States, if the frequency of launches is increasing, it is not because the public wants to change its perfume more often, but because, on both sides of the Atlantic, there is a lack of imagination, knowledge, authority. A lack of the requisite talents indeed - but also a lack of the ability to appreciate such talents as do exist. Too many perfumes launched too hastily, sometimes at too low a cost price, have no character of genuine worth. They are not go-ing to win strong and lasting commitment from the public. That is why these transitory fashions for second-rate products are something the public is quick to forget. Good products do not suffer this fate.

The reasons for this state of affairs are manifold. On the one hand, there are the self-confident promoters who chance their arm, keen to do well but lack ng any armory of expertise in composition and sampling by the nose. On the other there are the suppliers, who ought to know better but are unable to resist the demands of their customers. whom they themselves egg c11 with their repeated offers. Bu1 one cannot create "by numbers;" no way can these suppliers guarantee their customers a masterpiece every time.

So, all concerned go for more and more of these high-risk projects, thinking this will im-prove their chances. Both parties figure that they will make at least as much money with a large number of ephemeral formulations as they would with a few fragrances of real beauty which would go on selling over a long period- and which are in any event no longer to be had because true creativity has been muzzled. The question is whether this is the right sum to be doing.

Promotional budgets grow and grow. The question is how many budgets are not covered by returns for each of the few that are- not always thanks to a product that may be of questionable merit, but to the house's repute, its bosses' connections, a good PR man, appeal to the snob value of novelty, or whatever. Two or three launches which do not get off the ground are enough to drag a firm down if not to sink it- likewise enough to harm the supplier's reputatiol and to restrict the scope of his experinentatlon as failure is heaped on failure.

What is more, if a supplier attracts customers by offering formulations too cheap, he may indeed increase turnover but to the detriment of his net results, which may take a dangerous turn for the worse. This will result in the recruitr1ent of cut-rate per-fumers, novices or indifferent performers that is, who will only be able to produce worse and worse fragrances, which will get harder and harder to sell and shorter and shorter livec. The circle duly closes with the collapse of those who, armed with knowledge but little foresight. opt for a fast buck rather than tor plying their trade properly.

Laid bare and objectively examined, this is the foundation of the policy currently being operated by some people who, unfortunate-ly, have a considerable say. II is blind policy, with no deal and no future. The public, most of all the French public, will end up by doing without and turning its back on the whole business. That will be the moment to make a fresh start by offering the public fine perfumes-and that is why we must go on training people who are capable of doing just that. Likewise, that is why we must sus-tain the taste of both sides of the industry, and of the public, by seeing to it that, despite everything, a smell of lovely shape sees the Iight of day every now and then - to brighten a sky heavy with clouds.

After the Second International Paris Meetings

The Report of these Meetings and the press coverage cast Iight on what was said above and call tor some further remarks. Thus, we are told that two delegates denounced women's faithfulness to their perfumes as an outmoded relic of the past. The fact is that the faithfulness of a woman to her per-fume is very simply explained by the in-herent worth of the perfume. It has the qualities needed to please her-· to attract her in the first instance, and then to hold her. She does not get tired of it because it is beautiful, any more than I get tired of the countryside I have gazed over for 37 years because it is as lovely as ever. But I doubt there are many women who would stay faithful for long to a mediocre perfume. If they change, it is because the perfume is not a good one; so, they have good reason to show it the door.

One may liken perfume to a dress and change it just as often. But to wear a different dress every day you need a deal of money. To wear a different perfume every day you Iikewise need a deal of money. What price, then, the desire lo democratl ze perfume. As to the desire to trivialize it, that just shows a total lack of taste.

This point was thoroughly appreciated and strongly brought out by many delegates. Mr. Robert Ricci 1n particular gave vent to a notably sound reaction: "Perfume," he said, "is not, alas, just a piece of merchandise. Too many perfumers have no real ethic. The American woman tends to buy a perfume like she chooses a belt or a brooch. At the moment we are living in a cacophony of smells which confuses the senses and hardens resistance -whence the compul-sion to have stronger perfumes, because people are less and less good at using their sense of smell."

At the beginning of October 1985 I was invited to Westphalia to attend the Congress of German academics specializing in the Romance languages. A lady professor from Berlin University- she can't have been 40--- said this to me: "I have been using Femme for 25 years and I've never found another perfume I liked as much." I was astonished that she had managed to take to it so young, and she was very touched to find herself talking to its creator.

The way great perfumes form a bond with those who wear them is no reason to be sur-prised that trite ones do not hold their customers. But whether making sure it will hold women is a good criterion for a perfume is another thing again. Personally, I fell no temptation to make this my aim.

During these "round tables," it was disheartening to find that women, even distinguished ones, did not go so far as to place perfume on the artistic level of its for-mulation or the creation of its shape. Rather they relegated it to the ranks of disposable objects, to be thrown away after use like a blunt razor.

This attitude can be explained, I guess, by the current saturation of the market, where too few fine perfumes are to be found, and these are outnumbered to the point of domination by scented mixtures which lack character or have not been well designed. Women now between 20 and 50 were un-born, or at most 10 years old, in 1945. So they have never known the open market where only fine perfumes flourished. Their taste has been formed not by these but by what came after them. That is why they do not always respect the new market.

One might well ask the lady who treats perfumes like a basic used razor whether. after listening to a recording of Debussy's Clair de Lune. she feels inclined to throw the disk in the wastepaper basket. Once she realizes that the greatest perfumes are as precious as that disk, she is bound to respect them. The best way to convince her would be to create a really lovely perfume with a shape that pleases her. 1 guess she would stay true to it and press her friends to do the same.

We need to do something about reeducating these ladies, making them appreciate what a real perfume composition is, and teaching their noses and their minds to distinguish between that and all the "nice smelling" or "strong smelling" mixtures with which our shop windows are stuffed - and then to make these women of heightened sensibili-ty seek after something different from the kind of chattel they are so ready to scorn.

But to lure them away from their "hail and farewell" attitude, we first have to create some beautiful new shapes. The problem 1s that the present situation, blocking as it does all possibility of real creativity, stops us doing this. Why, you may ask. Because tor a certain number of today's bosses perfume is not a work of art meriting respect; they can-not allow it to become any such thing. It has to be reduced to an item of merchandise, so that they can suck the last drop of profit from it without remorse.

Given the marketing techniques which these practices have had a hand in developing, and the lack of experience in matters of composition of the new masters of perfume the world over, the exercise of real, free creativity has become next to impossible for the simple reason that judges truly capable of appreciating it are becoming few and far between.

As a result, the training of new "composers" tends to be built on soil which does not have a deal in common with the imaginative search for the great perfumes or with the talent this called for. By suppressing its func-tion, you make the organ atrophy.

The creation of original formulations, at once interesting and harmonious. demands deep hard-won knowledge, talent, and time. How many young perfumers receive this in-dispensable training nowadays? How many people are there who can give it to them?

There are of course potential talents around. But nobody gives them the time or the means to express themselves freely, still less the chance to blossom forth. The situa-tion is indeed a dramatic one; but many of the actors in the piece clearly remain divorced from it.

This is a vicious so are indeed with few talents free to exercise themselves and few prospects 'er I hose who can. From 1t springs the dominance of :rite formulations with a Precarious f uture, and thence the absence of the models or new talents to cut their teeth on.

This is the blockage that has to be cleared, by opening the eyes, cost what it may, of the tycoons and by showing them where their true long term interest lies - the critical fac-tor, that is, for the future and the perpetua-tion of their house. The great French houses are first in line for this treatment, for it is there, needless to say, that the bulk of poten-tial talent is to be found. Our foreign colleagues too could not but welcome with open arms an atmosphere of greater free-dom and understanding in which beautiful new perfumes could flourish.

There is something striking, something highly significant, about the promotional ar-ticles and all the various texts which are be-ing written about perfume today. The writers tend to spread themselves over the beauty of the flasks rather than of their contents. True. the flasks are otten very beautiful and original; their designers and manufacturers show plenty of talent, and manage to come up with something fresh despite the plethora of articles they churn out. But if PR and advertising persons lay such stress on the container rather than the contents, may it not be because tley are embarrassed to talk about tle latter? Maybe they lack the means to Judge the perfume objectively; maybe ever they find it disappointing.

Maybe they latch onto the presentation because they have doubts about the perfume. One has to ask what has made them go this way.

We are fishing in deep waters here. The need is to change a whole climate of opin-ion, a whole philosophy, I would go so far as to say a whole professional attitude. For if what we do 1s to put this, that or the other into the most elegant of flasks, we are not per-fumers any more. The public contents itself with buying just high-priced flasks; and hav-ing no reason to suppose that the contents of the "refills" will be any better than that of the original, they do not buy any more. One day a great creator of flasks confided to me his disgust at what finally got put into his creations.

The Report on the Meetings makes play with the alleged American fashion of impulse buying of small flasks. This is nothing more nor less than the exploitation of an ill-informed middle and down market clientele-something which gives me to recall with anger the arguments I had to do battle against some years back. All this has nothing whatever to do with the American clientele for No 5, Shalimar, L'Air du Temps and the like.

The fact of the matter is that they divide into miniflasks a series of suspect perfumes in which they have little faith, knowing that their mistrustful customers will try one after the other before rejecting them. Even if one could find one, there is little incentive to launch a great creation into this melee, for fear it gets lost in the wash, or damned by a world in which nobody knows anything about anything any more. The vicious spiral closes to a circle with the disappearance of a fine profession of which we used, despite all, to be proud.

This business of putting perfume in miniflasks might ostensibly be justified if they contained excellent products or if they had a really low cost. But we all know that, even with automatic bottling, small capacities attract high labor costs and overheads, and that the perfume in a small flask costs the buyer proportionately far more than that in a large flask. So all that is a delusion.

Some of the discussions in the Meetings blatantly mixed up two quite different activities - the creation of great perfume, and the manufacture of perfume products for the mass market. These are different in kind and do not mix. Gathering round the same table people who are not interested in the same clientele, who do not sell com-parable goods, who do not share the same problems, and thus who do not speak the same language, adds up to a misunder-standing about to happen. For they have nothing valid to say to one another. If each sticks to its last, there is no reason for con-flict between these two groups. They are complementary. They get the same materi-als from the same suppliers, but put them to different uses. No way does the world of fine perfume injure the mass market. Far from it, it provides the mass market with fine models and brings it customers who are interested in perfume but cannot afford a great one.

Sometimes the opposite happens Weighing off quantity against quality is no way to go about things; it never did anybody any good.

When one is sitting round a table, time virtually prevents one from tackling the fun-damental problems which stern from a technology so complex and so secret as the composition of perfumes. One cannot raise delicate questions in brief, desperately superficial exchanges in which little can really be said. To get results it would be worth providing a round table discussion on composition with discussion paper, a sound technical expose of adequate depth with copies circulated to participants for ad-vanced study and reflection.

In his latest book, L'hommen peril. the great ethologist Konrad Lorenz shows that man has turned onto a road to disaster which he calls "intraspecific selection" This leads the various human groups "t.o base their behavior on a one and the same single model and to put all their efforts into com-peting within that model." Surely this offers a striking analogy with what we find in one profession today-of the way perfumers, spellbound by a single more or less deserv-ed commercial success, are content to base their action on this success, ephemeral as it may prove to be. Lack of imagination and in-tuition, and refusal to accept hard work and responsibility are fatal diseases, for they diverge from man's true characteristics.

Nonetheless I remain convinced that suppliers, sound people aware of their responsibilities and men of tradition, wo have laboriously developed some interesting specialties or some tine new fragrances, will not stand for seeing these plunge to a sudden death. Rather they will try to sell them over a long period and to safeguard those that merit it. I would invite them most warmly to unite their efforts with ours to revive the profession and restore its morale; to change its standards and its climate; to reawaken in it ambition for its lost status; and to point it towards its destiny, which is to open paths, not to close them.

Some folks feel that life doesn't make sense. This drives them to seek instant pleasures, which will progressively lose their savor as time goes by. Yet life takes on a meaning worthy of commitment if one takes as one's goal the pursuit of beauty. Beauty in all its forms, of course- the beauty of soul, of body, of mind; the beauty of work wall done, however modest. After au, what stronger motive can there be than the prospect of achieving one's goal and the satisfaction of having done so? What could be more splen-did than the beauty of a face, the beauty of a woman, the beauty of nature in all its diversi-ty, the beauty of a timely and effective sport-ing gesture? Or the beauty of all the works of art, Earth's most enduring inhabitants? Or again the beauty of science which exalts the scientist and sometimes makes him shed human pettinesses?

In effect all these kinds of beauty merge into one, which is in turn the foundation of our ethic. Whatever name you like to call it by, this ethic is the search for beauty in all its aspects. That is why one should base all teaching, all training, all education in fact, on aesthetics. See to it that whatever you enterprise, whatever you accomplish, whatever you think, is beautiful. That way you will discover the secret of real success, the recipe tor relative happiness, and the key to freedom.

Like Plato's, my conception of beauty has its root in idealist thought; but its fulfillment rests on some extremely concrete criteria. In the free translation of Banquet which I in-cluded in l'Esthetique en question (PUF, 1977, page 34), I came to the very clear con-clusion that we must set our sights very high to provide ourselves with a stimulating aim, but at the same time we must keep both feet firmly on the ground, and remain in contact with the material, so as to get to know it well and be able to give it shape. Thus ac-complishing the work conjoins the idea and the sources of sensory consciousness, for -though Plato denies this- the percep-tion of beauty must necessarily have its origin in sensation.

It is because we have neglected to start by making thoroughgoing aesthetes of them that our philosophers, our teachers, our politicians, our engineers, our top managers all too often lack sureness of judgment when it comes to choosing their goals and making better choices about ways of achieving these. If he could really see the difference between the beautiful and the ugly, no way would man ever tolerate ugliness. And he would more often make a success of his life.

It is of this that we must convince the managers of the perfume world to make the best of them fully aware of the risk of ugliness each choice entails; to make them shun ugliness at all costs; and to make them aspire with all their strength to a single goal- fragrances with the beauty of shape that people of taste are impatiently waiting for.

Once they were more strongly m0tivated by the ques! for beauty, and better aware of what is beautiful and what 1s not, these top managers would be less easily taken in by charlatans. It is authentic and beautiful for-mulations, not nine days wonders, that will get them somewhere, allow them to be unreservedly proud of themselves, and en-sure that their efforts leave an enduring legacy.

Money and honors will not follow us into the next world, if there is one- still less so if there isn't. But the consciousness of beauty accomplished in our lives will brighten our last smile; for this and no other is the pillar which marks the victory we have won over ourselves.

Some advice to young perfumers

I can sum it up under four heads.

- Put your back into it. You can't take short cuts with effort Intense sustained effort is the only way to get results. The faint-hearted are few and far between in our profession.

Observe, observe, observe. Then ponder and force yourself to draw the right conclusions. Smell is easy and quickly learned. What demands exceptional qualities is far less the receptor system itself than the knack of interpreting the information it offers and putting this to good use.

The prerequisite for any attempt at composition is knowing how to place each piece of olfactory data precisely within one's men-tal repertoire, so that it does not get mixed up with anything else; to catalog it with such accuracy that the day of the memory will en-sure that what comes out is the smell in question and not any other. This olfactory accuracy is the test of sound initial judg-ment. His card index: of oltactory impressions will become the perfumer's vademecum: he will consult it and add to it every day of his life.

Having cleared the olfactory ground, you must start to form your general taste and Judgment. Cultivating your taste calls for in-formed contact with all aspects of beauty. You must constantly enrich it by critical study of the harmony of all kinds of shapes, so as to accomplish in your mind a (per-sonal) synthesis of all beautiful shapes-a synthesis which will become for you a universal reference, a benchmark, a leading light, a general principal of good taste. This will enable you to judge more surely what is beautiful and what is not.

Inevitably, this personal synthesis of taste is built up through your individual tempera-ment. Thus, tar from changing your per-sonality in any way, it respects the depths of your nature. through which it will grow richer and richer, and affirms your personality.

Under the influence of training in aesthetic appreciation. an aesthetic conscience will become second nature to you. It will safeguard you from many errors, become the thrust line of your thinking, a vigilant and exacting guide in the design of any kind of shape. This basic training holds good for any artistic career, indeed for any honest man who seeks to become cultured.

For culture is not, as people say, what is left after you have forgotten everything. When you have forgotten everything, what is left is nothing. Culture is the happy complex per-sonal synthesis of a very large number of pieces of knowledge which have first been well assimilated -- in other words, clearly arranged.

Along with all this the young perfumer will have followed the apprenticeship of his trade - the study of mixtures, affinities, antagonisms-and will have carefully noted and assimilated all the observations involv-ed in this. Armed with this and his cultural training, he will not be at a loss when it comes to composing a perfume. The educa-tion and refinement of his taste, the aesthetic forming of his thinking. the broad knowledge of shapes he will have acquired-these will lead him quite naturally to imagine olfactory shapes. All this will serve to trigger in him associations of ideas which need to be realized in a perfume. He will think in combinations of smells as naturally as others do in music or in painting.

This and no less, my young friends, is the price you have to pay to become true com-posers, to have good ideas for perfumes without becoming indebted to others or to machines which poach their formulas. Let's hope you can take all that in and thus arm your will for a lifelong contest for there is no retreat for those in love with their work

AN APPEAL TO PERFUMERS

The problem

A canvas covered in aimless splodges of color may find some collector who will take ii for a picture. Given certain ground rules, some notes sprinkled at random over the stave may produce something that passes tor a melody. Likewise, any mixture of aroma chemicals, tossed together without coherence or inspiration, is bound to have an outside chance of acquiring some smell capable of pleasing someone.

But a great many laymen may be wrong, and the resulting figures may create an illusion.

DEFINITION

So what is a perfume?

It is a "composition" of different fragrant products diluted in ethyl alcohol. 11 is not just a mixture, because before you mix you must crea1e. To create is first to conceive of an original "olfactive shape" which is original and from this abstract conception laborious-ly work out a compounding formula which will be the perfume's scenario. As the constit-uent parts of the formula are combined, the olfactive concept takes on a concrete form.

For this composition to have aesthetic merit, sufficient of its constituent parts must have been deliberately chosen and they must be proportioned in such a way that their com-bination gives a specific and therefore, recognizable olfactive shape). That shape should be both interesting and harmonious.

It is this which makes a mixture into a perfume and a perfume into a work of art. So this mixture is lot only a· ·scent:" the pleasant sensation received from perfume is not only a 'plus value" over and above is essential function, which is to forge a link. with the wearer, 1n harmony both physically and psychologically.

That is something which many women understand very clearly. Through their love of perfume they realize that perfume is something aesthetic, something whose beauty should be in harmony with their own.

History

For French perfume, the zenith was reached between 1905 and 1940, thanks to a handful of great creative geniuses: Guerlain, Houbigant, Coty, Caron, Chanel and Lanvin just to mention a few major names.

In those days, the head of most perfume houses was perfumer, manager and salesman combined; in other words, he was both connoisseur and businessman. Above all, he was an artist, which explains the quality of the perfumes and their worldwide impact The entry of the great couturiers, themselves artists, into the world of perfume prolonged the golden age until after the war.

Today, the bosses in many more houses are first class businessmen, college graduates. but strangers to the metier of a perfumer and its disciplines. Once regarded as works of art. perfumes are now sold by methods which have been tried and proven 1n the US, but in the vastly different world of industry, this has progressively led the marketing men to regard them, not as works of art, but as industrial products, governed by the same operating rules, They have lost sight of the tact that, first and foremost, perfume depends on taste, not science, nor tricks, still less chance.

When the heads of the houses no longer created the perfumes and delegated this task to paid staff and suppliers who were au fait with modern sales methods but not qualified to make value Judgments on per-fume, perfume came to be stripped of its mystery and its nature changed. The climate for free and fertile creation no longer ex-isted. So there was a switch towards the mass production of scented mixtures, just a little too often using methods dominated by mechanical considerations, where physical features triumph over aesthetics and over the shape and its harmony.

Thus the training of new perfumers is organized on different lines than those which artistic creativity demands. The results are edifying; of more than 100 per-fumes launched in France between 1966 and 1983, less than 10 have made a name for themselves. In other words, over 90 falled.

This shows the poor quality of those products, it's true, but it also shows the public are less easily fooled than the marketing men.

Foreign perfumes have been able to break into the French market, less because of their owl intrinsic merit than as a result of the decline 1n French perfumes, which have too often been mediocre and sometimes frankly awful. It is also because the experienced cl1entele in France is growing old and gradually vanishing, and their successors have no sure benchmarks to go by•-- any more than the now leaders of the profession do.

It's worth noting that, only one of today's top men in the business lived through the 1925 -1940 period when all the master-pieces were still there. Today's men have never smelt these masterpieces and, since they are incapable of imagining them, they haven't even an inkling of what really great perfumes were like. When you talk to them about it, it's as if you were conjuring up history which cannot be reconstructed. So they believe, in alt good faith, that the suc-cesses of the present-day perfume industry represent a scale of values to be aimed at, and they have no yardstick against which they can measure and judge progress.

Besides. it is abundantly clear that the cur-rent general trend of most houses is to blame all the problems on marketing and to agonize over that, rather than over the ac-tual products they sell, because they are none too sure how to evaluate them.

So. it seems, we are to see so-called "luxury" perfumes launched, selling at knock-down prices. Since the age of miracles is past and since some costs just cannot be cut, there is only one answer; either the perfumes will not be all they are cracked up to be, or profit margins will be reduced to a point where it amounts to "dumping." If others in turn allow themselves to get caught up in this trend, then the overall quality of perfume will go down another notch. It really doesn't need to do that.

Artistic directors

The failures of French perfume cannot be at-tributed solely to the gap left by the pro-prietor-perfumer-connoisseurs of old. This is a gap which can only be filled- unless new proprietors of equal genius can be found - by an artistic director carefully train-ed for just this purpose. He will be chosen (without insisting on paper qualifications. tor no amount of schooling can instill intuition) primarily to his cultural qualities, for the evidence he can give of his good taste in various spheres, and for his olfactory ap-titude. Above all, and this is the essential point, he will have to be a man of taste; and they don't grow on trees.

By successive stages, moving from depart-ment to department and going to retailers too, after a long initiation in compounding (ISIPCA and suppliers), he will totally familiarize himself with perfume, how to make it, present it and judge it. Thus prepared and educated, he will be able to collaborate intelligently and effectively with the in-house perfumer, if there is one, or with those of suppliers. If he has to deal with true creative artists, he will do his best to under-stand them, to discuss things tactfully but without interfering in such a way as to upset the researchers and prejudice their common interest. But once he has won his spurs, he must be able to tell them what he wants and they will be under his general direction.

The perfumes will be free to carry out his studies in quiet collaboration with the artist or director; they will give each other mutual support. When the time comes the artistic director will assume his responsibiilties and act as decision-maker. Both the artistic director's position and the perfumer's future must depend on the success or failure of their work; the two must build up a fruitful and friendly interdependence, the guarantee of future progress and prosperity for the house.

It is only by creating this new breed of executive that the world perfume industry will, together with a re-educated public, once again make royal progress and it will be seen that there is more profit in beauty than in ugliness. The form of this training will need honing down by trial and error until the right formula has been found; but once it has been found, what wonderful things there are in store for young people of taste! It is no more and no less than a school for critics. of course, it goes without saying that an artistic director need not be a man - tar from it.

The return to creativity

To put creativity back in its rightful place, we need to attract the right people, people who will take up the challenge. First we must bring them out of anonymity if we want to keep them and motivate them to give of their best If it is fair to publish the name of the creator of a pretty 'task, we should not forget the name of 1he man who created the concerns. this can only be done it imposters and creations not worthy of the name are rigorously excluded. Otherwise, re-birth is out of the question.

The true masterpiece must bear the master's signature and be treated accord-ingly; it must be subject to originator's rights and given the artistic protection afforded by the French law of March 11, 1957. This law does apply to great perfumes, as was formally established by "Le Droit de Par-fl1m" by J.P. Pamoukdjian.

To be perfectly fair, the artistic rights ought to be shared between the creator of the per-fume and the house that shelters him under its flag, bearing as it does enormous risks and responsibilities.

The situation will then be crystal clear; we shall know who succeeds and who fails. Houses will have to take up the cudgels if great perfumes are copied and vigorously pursue counterfeits. This should surely clear the decks.

If a competent artistic director gets involved in creation as well as promotion, creation will flourish and great new creators will rise up again. Today there are many more people making perfumes than once there were; among them there have to be some mighty talents. but they are kept straight-jacketed by briefs" which are often illusory, some- times puerile and without any real signif-icance; there must even be people who are pulled through the hoop if they have an original idea which baffles the laymen. In the old days, no way would anyone have dared to take the proprietor-perfumer to task! That is why the grandes marques were able to make the grade. A talent allowed to go rusty dries up, so those who, due to ig-norance or blindness have either brought about this impasse or just let it happen, have a lot to answer for.

To break down the impasse we must at all costs open the eyes of those (and there are no longer many of them) who are still capable today of promoting great perfumes. We must show them that true success is not a matter of luck; it is based on the sure value of authentic quality, which is born of talent. To recognize talent, to appreciate it and not confuse it with a charlatan, takes experience and taste - and they cost money, they can-not be improvised or circumvented without risk. Exploiting snobbery may make money, but it cannot make masterpieces. Master-pieces bring other rewards; they make money too, but in a different way and over a longer period.

To get over this present sickness, we must again let talent have its head. we must nur-ture it and respect it. If the last remaining promoters of beauty are unwilling to listen to this appeal or if they do not respond, if they cannot recognize their last chance, the great French perfume industry will finally be con-demned to pass into history.